FLORA OF TROPICAL EAST AFRICA

ESCALLONIACEAE

B. Verdcourt

Trees or shrubs. Leaves simple, alternate, rarely subopposite or sub-verticillate, usually glandular-serrate; stipules usually absent or minute. Flowers hermaphrodite or less often dioecious or polygamous, mostly in terminal or axillary racemes, panicles, or cymes; in one genus epiphyllous. Sepals 4–5, mostly united at the base or rarely free, imbricate or valvate, often persistent. Petals 4–5, free or rarely connate into a short tube, imbricate or valvate. Disk annular or with lobes alternating with the stamens. Stamens (4–)5(–6), sometimes alternating with staminodes, perigynous, free; anthers 2-celled, opening by longitudinal slits. Ovary superior or inferior, syncarpous or apocarpous, 1–6-locular; ovules with axile or parietal placentation; ovules numerous; styles 1–6, free or ± joined. Fruit a capsule or berry. Seeds few to many, with small or large embryo and copious endosperm.

A rather small somewhat poorly defined family, formerly included in Saxifragaceae, with about 23 genera when defined in a broad sense but no more than 7 when dis-membered into smaller families. For the purposes of the Flora Brexiaceae (*Brexia*) and Montiniaceae (*Grevea*) have been considered worthy of separate family status; the remaining genus, *Choristylis*, is retained in Escalloniaceae although it is sometimes separated, together with *Itea*, as a small family Iteaceae* mainly characterized by extraordinary 2-porate subisopolar pollen grains.

Various species and hybrids of *Escallonia* are grown quite frequently in gardens in Nairobi and elsewhere, e.g. *E. punctata* DC. (*E. rubra* (Ruiz & Pav.) Pers. var. *punctata* (DC.) Hook. f.), Kenya, Kiambu District, Muguga, Hort. Greenway, 31 Mar. 1963, *Greenway* 10878!

CHORISTYLIS

Harv. in Hook., Lond. Journ. Bot. 1: 19 (1842); V.E. 3(1): 287 (1915); E. & P. Pf., ed. 2, 18a: 214 (1930); Erdtman in Grana Palynologica 1: 3 (1954) & in Webbia 11: 407 (1955); G.F.P. 2: 31 (1967)

Shrubs. Leaves alternate, closely glandular-serrate, penninerved; stipules minute, linear. Flowers hermaphrodite or polygamous, small, in much-branched axillary panicles shorter than the leaves. Calyx-tube obconic, adnate to the ovary; lobes 5, subulate, distant, triangular at the base, puberulous, persistent. Petals 5, ovate-deltoid, perigynous, broad at the base and confluent with the epigynous disk, valvate, puberulous, persistent. Stamens 5, alternating with the petals, inserted at the margin of the disk; filaments subulate; anthers dorsifixed, hairy, small, ovoid, the cells separated by a thick connective. Ovary partly inferior, 2-locular; ovules numerous on axile placentas; styles 2, subulate, adherent or diverging; stigmas capitate. Capsule half-superior, 2-locular, dehiscing septicidally between the styles, many-seeded. Seeds irregularly obovoid or oblong-obovoid, slightly curved with a thick tesselated testa.

---

* E.g. in Willis " A Dictionary of Flowering Plants and Ferns ", ed. 7: 584 (1966), where a very narrow view of the family Escalloniaceae is taken.

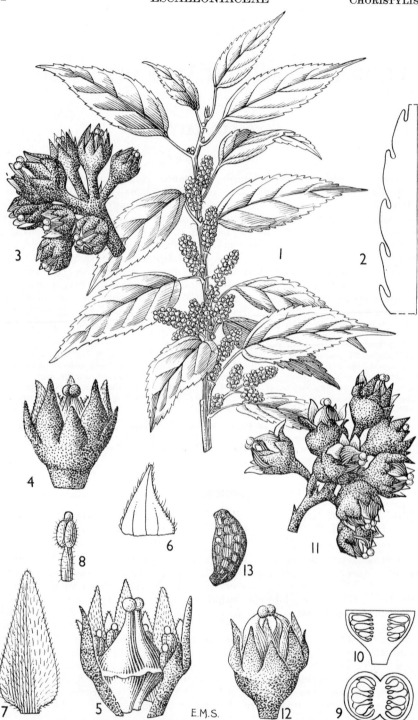

FIG. 1. *CHORISTYLIS RHAMNOIDES*—**1**, habit, × ⅔; **2**, outline of leaf-margin, × 6; **3**, inflorescence, × 6; **4**, flower, ×12; **5**, same, opened out to show ovary, ×12; **6**, sepal, ×24; **7**, petal, × 24; **8**, stamen, × 24; **9**, diagrammatic transverse section of ovary; **10**, diagrammatic longitudinal section of ovary; **11**, infructescence, × 4; **12**, fruit, × 6; **13**, seed, × 24. 1–10, from *Clements* 79; 11–13, from *Buchanan* 1432. Drawn by Miss E. M. Stones.

A monospecific genus restricted to E., Central and S. Africa and which presumably migrated northwards along the mountains, but seems to be absent from the more recent volcanic mountains in E. Africa.

C. rhamnoides *Harv.* in Hook., Lond. Journ. Bot. 1: 19 (1842) & in Fl. Cap. 2: 308 (1862); Brenan in Mem. N.Y. Bot. Gard. 8: 432 (1954); Boutique in B.J.B.B. 34: 504 (1964); Liben in Fl. Congo, Saxifragaceae: 2, t. 1 (1969). Type: South Africa, Cape Province, Katberg, *Brownlee* (TCD, holo., K, iso.!)

More or less erect or climbing shrub 1·8–7 m. tall or long; stems slender, diverging from the base, the young shoots often purplish, puberulous. Leaves petiolate; blades ovate or oblong to ovate-lanceolate, 1·5–10·3(–12) cm. long, 0·9–5·5 cm. wide, acute to acuminate or rarely rounded at the apex, cuneate to unequally rounded at the base, glabrous or with sparse hairs on the main nerves beneath and often with domatia, somewhat glossy above, margins sharply and closely serrate; costa and main nerves impressed above, prominent beneath, the tertiary venation closely reticulate; petioles often purple, 0·5–2 cm. long. Inflorescences 1–4·5 cm. long, 1–2·5 cm. wide, many-flowered, finely pubescent-tomentose, the flowers creamy white or greenish yellow, sweetly scented; peduncles 0–1·2 cm. long; pedicels 1–2(–5) mm. long; bracts and bracteoles narrowly triangular, 0·5–1·5 mm. long, 0·2–0·5 mm. wide. Calyx-tube 1–1·5 mm. long, 1–1·5 mm. wide; lobes 0·5–2 mm. long, 0·3–0·5(–1) mm. wide at the base. Petals (1·6–)2–3·3 mm. long, 1–2 mm. wide at the base, 3-nerved. Styles yellowish, turning brown, 0·5–1 mm. long; stigmas 0·3–0·5 mm. in diameter, free or sometimes cohering for a time. Capsule campanulate to turbinate below, conical above, 3–5(–6) mm. long, 3 mm. wide, finely pubescent, slightly ribbed. Seeds brownish, ± 1 mm. long, 0·5 mm. wide.

UGANDA. Kigezi District: Kanaba, Luhizha, Mar. 1947, *Purseglove* 2363!
TANGANYIKA. Lushoto District: W. Usambara Mts., Shume, World's View, 6 Nov. 1947, *Brenan & Greenway* 8297!; Mbeya Peak Forest Reserve, 21 May 1958, *Gaetan Myembe* 35!; Songea District: Matengo Hills, Mpapa, 6 Oct. 1956, *Semsei* 2515! & same area, Mtete, 13 Sept. 1936, *Zimmer* 80!
DISTR. **U**2; **T**3, 6 (see note), 7, 8; Burundi, Zaire, Mozambique, Malawi, Rhodesia, South Africa (Transvaal, Natal, Cape Province), Lesotho and Swaziland
HAB. Evergreen forest and forest edges; 1500–2400 m.

SYN. *C. shirensis* Bak. f.* in Trans. Linn. Soc., ser. 2, Bot. 4: 13, t. 3/1–6 (1894); T.T.C.L.: 196 (1949). Types: Malawi, Mt. Mlanje [Milanji], *Whyte* 53 & Shire Highlands, *Buchanan* 158 & 1468 (all K, syn.!, BM, isosyn.!)
  ? *C. ulugurensis* Mildbr. in N.B.G.B. 12: 191 (1934); T.T.C.L.: 196 (1949). Type: Tanganyika, Uluguru Mts., S. slope of Lukwangule peak, *Schlieben* 4293 (B, holo. †)

NOTE. No material has been seen from the Uluguru Mts., but it seems likely that *C. ulugurensis* is merely a form of *C. rhamnoides* as suggested by Liben (*loc. cit.*: 1 (1969)). He found no isotypes. Mildbraed gives the leaf dimensions as 8–12 × 3–4 cm. and the inflorescence dimensions as 10 × 8 cm. which is in excess of any specimens seen; he also states that only flowers with sterile sessile anthers were seen. Topotypic material is needed. There is a tendency for the Tanganyika material to have larger leaves and capsules and shorter styles than in Central and S. African plants.

* A *C. virescens* Bak. f., referred to in the Index Kewensis as published on the same page, seems to be fictitious.

INDEX TO ESCALLONIACEAE